YOUR KNOWLEDGE HAS VALUE

- We will publish your bachelor's and master's thesis, essays and papers

- Your own eBook and book - sold worldwide in all relevant shops

- Earn money with each sale

Upload your text at www.GRIN.com and publish for free

Bibliographic information published by the German National Library:

The German National Library lists this publication in the National Bibliography; detailed bibliographic data are available on the Internet at http://dnb.dnb.de .

This book is copyright material and must not be copied, reproduced, transferred, distributed, leased, licensed or publicly performed or used in any way except as specifically permitted in writing by the publishers, as allowed under the terms and conditions under which it was purchased or as strictly permitted by applicable copyright law. Any unauthorized distribution or use of this text may be a direct infringement of the author s and publisher s rights and those responsible may be liable in law accordingly.

Imprint:

Copyright © 2019 GRIN Verlag
Print and binding: Books on Demand GmbH, Norderstedt Germany
ISBN: 9783346011312

This book at GRIN:

https://www.grin.com/document/496168

Evans Mogoi

Asymptotic Condition and Deficiency Indices of Difference Operators

GRIN Verlag

GRIN - Your knowledge has value

Since its foundation in 1998, GRIN has specialized in publishing academic texts by students, college teachers and other academics as e-book and printed book. The website www.grin.com is an ideal platform for presenting term papers, final papers, scientific essays, dissertations and specialist books.

Visit us on the internet:

http://www.grin.com/

http://www.facebook.com/grincom

http://www.twitter.com/grin_com

Asymptotic condition and Deficiency Indices of Difference Operators

N. M. Mogoi Evans

Department of Mathematics

Eregi Teachers Training College

Abstract

Jacobi matrices together with Sturm-Liouville operators and have already been developed in parallel for many years. However not much in terms of spectral theory has been done in the discrete setting compared to the continuous version especially in higher order operators. The main objective of this study is to compute the deficiency indices of Fourth order difference operator.

Introduction

In this study, we have investigated the deficiency indices of a fourth order self-adjoint extension operator of minimal operator generated by difference equation;

$$
\begin{aligned}
\mathcal{L}y(t) \;=\; & w^{-1}(t)\triangle^4 y(t-2) - i\{\triangle(q(t)\triangle^2 y(t-2)) + \\
& \triangle^2(q(t)\triangle y(t-1)\} - \triangle(p(t)\triangle y(t-1)) + \\
& i\{r(t)\triangle y(t-1) + \triangle(r(t)y(t)\} + m(t)y(t), \qquad (0.0.1)
\end{aligned}
$$

defined on a weighted Hilbert space $\ell_w^2(\mathbb{N})$ with the weight function $w(t) > 0$, $t \in \mathbb{N}$ where p(t), q(t), r(t) and m(t) are real-valued functions.Here the equation is in the form that makes it symmetric and also of order 4. In this case the coefficients are allowed to be unbounded. $\triangle$ is a forward difference operator such that $\triangle f(t) = f(t+1) - f(t)$; for $t \in \mathbb{N}$. The method applied is asymptotic summation as outlined in Levinson-Benzaid -Lutz theorem [5] and whose spectral parameter uniform version is given in [1, 2, 3]. For simplicity in computation and analysis ,we have assumed that $w(t) = 1$ unless otherwise stated. For the spectral analysis we have solved the equation $\mathrm{L}y(t) = zy(t)$ where L is the difference operator generated by (0.0.1) and z is the spectral parameter,$z \in \mathbb{C}$. These results has been an extension of some known spectral results of fourth order differential operators to difference setting.

Basic Concepts

Definition 0.1. A mapping $\triangle$ is known as forward difference operator if for any function $f(t), t \in \mathbb{N}$ then

$$\triangle f(t) = f(t+1) - f(t).$$

Similarly $\triangle^*$ or ∇ is backward operator if

$$\nabla f(t) = f(t) - f(t-1).$$

Definition 0.2. Let H be a separable Hilbert space and let T be a densely defined symmetric linear operator on H. The operator T is closed if its graph

$$\{x \oplus Tx \in H \oplus H : x \in D(T)\}$$

is closed. If T' is a symmetric operator on H with $D(T) \subset D(T')$ and $T' - \backslash D(T) = T$ we call T' a symmetric extension of T. symmetric operators have maximal symmetric extensions and the maximal symmetric extensions are closed only if not self-adjoint. In order to convert equation (0.0.1), into a first order system, we define the vector valued functions $x(t), u(t)$ and $y(t)$ by,

$$x(t) = (x_1(t), x_2(t))^{tr}, \; u(t) = (u_1(t), u_2(t))^{tr}, \; y(t) = (x(t), u(t))^{tr}$$

where the superscript tr denotes transpose and

$$
\begin{aligned}
x_1(t) &\quad- \quad y(t-1) \\
x_2(t) &= \triangle y(t-2) \\
u_1(t) &= p(t)\triangle y(t-1) - \triangle^3 y(t-2) + i\{\triangle(q(t)\triangle y(t-1)) + \\
&\quad q(t)\triangle^2 y(t-2)\} - ir(t)y(t) \\
u_2(t) &= \triangle^2 y(t-2) - iq(t)\triangle y(t-1).
\end{aligned}
$$

Now we let

$$x(t) = \left(\begin{array}{c} x_1(t) \\ x_2(t) \end{array} \right)$$

and

$$u(t) = \left(\begin{array}{c} u_1(t) \\ u_2(t) \end{array} \right)$$

Therefore the discrete linear Hamiltonian system as outlined by Hinton and Shaw[6] for differential operators and discritised by Shi[7] is of the form

$$J \triangle Y(t) = [zW(t) + P(t)]R(Y)(t) \tag{0.0.2}$$

where $t \in \mathbb{N}$, $W(t)$ and $P(t)$ are 4 x 4 complex Hamiltonian matrices. $W(t) = \text{diag}(w(t), 0..., 0)$, $w(t)$ is a weighted function, $x(t), u(t) \in \mathbb{C}^2$, J is a symplectic matrix, that is

$$J = \begin{pmatrix} 0 & -I_2 \\ I_2 & 0 \end{pmatrix} \quad \text{and} \quad P(t) = \begin{pmatrix} -C(t) & A^*(t) \\ A(t) & B(t) \end{pmatrix}.$$

For non-zero elements of 2 x 2 matrices A,B, and C are given by

$$A_{1,2} = 1, A_{2,2} = iq, B_{2,2} = 1, C_{1,1} = m, C_{1,2} = -C_{2,1} = ir, \text{ and } C_{2,2} = p$$

Definition 0.3. The deficiency indices of the operator L is the pair (N_-, N_+) defined by, $\dim N(L^* \pm iI)$ and denoted by N_- and N_+, for $\dim N(L^* - i)$ and $\dim N(L^* + i)$ respectively. Here $N(L^* \pm i)$ is the null space of $L^* \pm i$. Thus $2 \leq N_-, N_+ \leq 4$ and if $N_- = N_+$ then, there exists a symmetric self-adjoint extension H of an operator L.

Definition 0.4. Let

$$Y_\alpha(., z) = \begin{pmatrix} U_\alpha(., z) \\ V_\alpha(., z) \end{pmatrix}$$

be the fundamental matrix of

$$J \triangle \begin{pmatrix} x(t) \\ u(t) \end{pmatrix} = \begin{pmatrix} 0 & -\triangle \\ \triangle & 0 \end{pmatrix} \begin{pmatrix} x(t) \\ u(t) \end{pmatrix} = \begin{pmatrix} -C(t) + zW & A^*(t) \\ A(t) & B(t) \end{pmatrix} \begin{pmatrix} x(t) \\ u(t) \end{pmatrix} \tag{0.0.3}$$

with initial values of

$$Y_\alpha(a, z) = \begin{bmatrix} \alpha_1^* & -\alpha_2^* \\ \alpha_2^* & \alpha_1^* \end{bmatrix},$$

where

$$\alpha_1, \alpha_2$$

satisfy

$$\alpha_1 \alpha_1^* + \alpha_2 \alpha_2^* = I_2, \alpha_1 \alpha_2^* - \alpha_2 \alpha_1^* = 0_2 \text{ and } \alpha_1 \alpha_2 Y(a) = 0 \tag{0.0.4}$$

α_1 and α_2 are 2 x 2 matrices, that is $\alpha = (\alpha_1, \alpha_2) \in \mathbb{C}^{2 \times 2}$.

U_α, V_α are 4 x 2 complex-valued matrices whose every column solves $Ly = zy$ and that $V_\alpha(,..z)$ satisfy self-adjoint boundary conditions at a. Thus, columns of $Y_\alpha(,...z)$ span the 4-dimensional vector space of solutions of (0.0.3). Therefore in the limit point case with $Imz > 0$ one has a matrix $M \in \mathbb{C}^{2 \times 2}$ such that

$$\mathcal{X}_\alpha(t, z) = Y_\alpha(a, z) = \begin{bmatrix} I_2 \\ M(z) \end{bmatrix} = U_\alpha(t, z) + V_\alpha(t, z)M(z)$$

where $\mathcal{X}_\alpha(t, z)$ satisfy the boundary conditions of (0.0.4). It has been shown in [7] that if L is limit point as $t \to \infty$, then one can construct the M-matrix M(z) for the Hamiltonian restriction to $[a, \infty)$ with Dirichlet boundary conditions. To do this, let

$\begin{pmatrix} W_1(a, z) \\ W_2(a, z) \end{pmatrix}$ be a system of 2 square summable solutions for $Imz > 0$.

Then from the theory of Hinton and Shaw [6], it follows that this solutions also arise from $Y_a(t, z) \begin{pmatrix} I_n \\ M(z) \end{pmatrix}$, where $Y_a(t, z)$ is the fundamental solution of the system satisfying the appropriate boundary conditions at a.

Hamiltonian System

In order to define discrete Hamiltonian system of (0.0.1), we need to introduce quasi-differences as explained in [7] and [3, 4]. Thus we define the vector valued functions $x(t), u(t)$ and $Y(t)$ as in chapter one. Therefore we introduce the spectral parameter $z, z \in \mathbb{C}$ and solve the equation

$$Ly(t) = zy(t) \tag{0.0.5}$$

In such a case, the Hamiltonian system (0.0.2) which can be rewritten as

$$J\Delta \begin{pmatrix} x(t) \\ u(t) \end{pmatrix} = \begin{pmatrix} 0 & -\Delta \\ \Delta & 0 \end{pmatrix} \begin{pmatrix} x(t) \\ u(t) \end{pmatrix} = \begin{pmatrix} -C(t) + zW & A^*(t) \\ A(t) & B(t) \end{pmatrix} \begin{pmatrix} x(t) \\ u(t) \end{pmatrix}$$

with A,B and C given by

$$A = \begin{bmatrix} 0 & 1 \\ 0 & iq \end{bmatrix}, \quad B = \begin{bmatrix} 0 & 0 \\ 0 & 1 \end{bmatrix}, \quad \text{and} \quad C = \begin{bmatrix} m - z & ir \\ -ir & p \end{bmatrix}$$

In order to ensure the existence, uniqueness and continuity of the solutions of initial value problem of (0.0.5), we need that $I_2 - A$ is invertible in $\mathbb{N}$, but this is always true for the fourth order case as long as $a > 2$.

Let $(I_2 - A)^{-1}$ be denoted by E, then in line with the analysis of shi[7], (0.0.5) then has a first order system of the form

$$Y((t+1), z) = S(t, z)Y(t, z) \qquad (0.0.6)$$

where

$$S(t, z) = \begin{bmatrix} E & EB \\ CE & I - A^* + CEB \end{bmatrix}$$

The 2x2 block matrices are then obtained from;

$$E = \begin{bmatrix} 1 & \frac{1}{1-iq} \\ 0 & \frac{1}{1-iq} \end{bmatrix}, \quad EB = \begin{bmatrix} 0 & \frac{1}{1-iq} \\ 0 & \frac{1}{1-iq} \end{bmatrix}$$

$$CE = \begin{bmatrix} m - z & \frac{m-z+ir}{1-iq} \\ ir & \frac{p-ir}{1-iq} \end{bmatrix}, \quad CEB = \begin{bmatrix} 0 & \frac{m-z+ir}{1-iq} \\ 0 & \frac{p-ir}{1-iq} \end{bmatrix}$$

Hence (0.0.6) becomes,

$$\begin{bmatrix} x(t+1, z) \\ u(t+1, z) \end{bmatrix} = S(t, z) \begin{bmatrix} x(t, z) \\ u(t, z) \end{bmatrix} \qquad (0.0.7)$$

where $S(t, z)$ is a 4 x 4 transfer matrix given by;

$$\begin{bmatrix} 1 & \frac{1}{1-iq} & 0 & \frac{1}{1-iq} \\ 0 & \frac{1}{1-iq} & 0 & 1-iq \\ m-z & \frac{m-z+ir}{1-iq} & 1 & \frac{m-z+ir}{1-iq} \\ -ir & \frac{p-ir}{1-iq} & -1 & \frac{1+q^2+P-ir}{1-iq} \end{bmatrix}$$

The system (0.0.6) is now solved using asymptotic summation. The spectral multiplicity is computed via M-matrix.

Asymptotic Summation

Levinson-Benzaid -Lutz theorem is useful in asymptotic summation since the results is the extension of Levinson's theorem from differential calculus to difference setting .This result first appeared in the paper of Benzaid and Lutz [5] and has been extended by many authors, namely, Behncke and Nyamwala [3, 4] as well as Shi [7].Thus asymptotic summation is based on Levinson-Benzaid-Lutz theorem . The statement of this theorem implies that we solve for the eigenvalues of the matrix $S(t, z)$. In such a case ,we determine the characteristic polynomial $\det(S(t,z) - \lambda I_4)$ which gives;

$$\mathcal{P}(t, \lambda, z) =$$
$$(1-\lambda)^2\left[\frac{1}{(1-iq)^2} + \frac{q^2}{(1-iq)^2} - \frac{2\lambda}{1-iq} - \frac{\lambda q^2}{1-iq} - \frac{\lambda p}{1-iq} + \frac{\lambda ir}{1-iq} + \lambda^2\right] - \frac{2ir\lambda(1-\lambda)}{1-iq} + \frac{\lambda^2 m}{1-iq}.$$

Thus multiplying $\mathcal{P}(t, \lambda, z)$ by $\frac{1-iq}{\lambda^2}$ so that if λ is a root, then $\overline{\lambda}^{-1}$ is also a root, we obtain;

$$F(t, \lambda, z) = [(1-\lambda^{-1})^2(1-\lambda)^2 + p(1-\lambda^{-1})(1-\lambda) + (m-z)] + [q(1-\lambda^{-1})(1-\lambda)(i\lambda + (i\lambda)^{-1}) + r(i\lambda + (i\lambda)^{-1})].$$

In order to have a polynomial of real coefficients, we apply a transformation $\lambda = \frac{is+1}{is-1}$ that maps upper half plane into the interior of a circle, such that

$$(1-\lambda^{-1})^2(1-\lambda)^2 \;=\; \frac{16}{(s^2+1)^2}$$

$$(1-\lambda)(1-\lambda^{-1}) \;=\; \frac{4}{(s^2+1)}$$

$$i\lambda + (i\lambda)^{-1} \;=\; \frac{4s}{s^2+1}$$

and

$$Q_0(s,t,z) = \frac{16}{(s^2+1)^2} + \frac{4p}{s^2+1} + (m-z) + \frac{4q}{s^2+1}\left(\frac{4s}{s^2+1}\right) + \frac{4rs}{s^2+1}$$

The terms in the denominator can be eliminated by multiplying through by $(s^2+1)^2$ so that we have

$$Q(s,t,z) = (s^2+1)^2 Q_0(s,t,z)$$

and is given by

$$Q(s,t,z) = ms^4 + 4rs^3 + (4p+2m)s^2 + (16q+4r)s + (16+4p+m). \quad (0.0.8)$$

Since the transformation of $(0.0.5)$ into Levinson-Benzaid -Lutz form by asymptotic summation involves diagonalisation, we need that the eigenvalues of $S(t,z)$ be distinct. By considering the resultant or the discriminant of $\mathcal{P}(\lambda, t, z)$ and $\partial_\lambda \rho(\lambda, t, z)$ one can show just like in [2], that there are only finitely many spectral values z for which $\mathcal{P}(\lambda, t, z)$ has multiple roots. Let $\omega_1 < \omega_2 < ... < \omega_k$ denote all of the real spectral values z

7

leading to multiple roots. Following [2], the analysis will be restricted to small complex neighborhoods of $z_0 \in (\omega_i, \omega_{i+1})$, $\quad i = 0, ..., k$ where $\omega_0 = -\infty$ and $\omega_{k+1} = \infty$. For a given $z_0 \in (\omega_i, \omega_{i+1})$, one can now choose $\epsilon > 0$ and $a > 0$ so that $\mathcal{P}(\lambda, t, z) = 0$ has no multiple roots for any z

$$z \in \mathcal{K}_\epsilon(z_0) = \{z | |z - z_0| \leq \epsilon, \quad Imz \geq 0\}$$

and $t \geq a$.This is possible because the roots of $\mathcal{P}(\lambda, t, z)$ depend analytically on the coefficients. Throughout the study, it may be necessary to adjust a and ϵ repeatedly.This will be done without mentioning.

Bounded Coefficient

Definition 0.5. The coefficients $q(t), r(t), p(t)$ and $m(t)$ are said to be almost constant coefficients if there exists constants c_q, c_r, c_p and c_m such that

$$q(t) \to c_q, r(t) \to c_r, p(t) \to c_p, \text{ and } m(t) \to c_m \text{ as } t \to \infty. \quad (0.0.9)$$

In this case,the coefficients $q(t), r(t), p(t)$ and $m(t)$ are bounded. With this assumption, we have the following theorem, which proves that in the case of bounded coefficients, then there exists an interval in which the singular continuous spectrum of H is absent.

Theorem 0.6. *Let H be self-adjoint extension operator of the minimal difference operator generated by (0.0.1). Assume the coefficients are almost constant, then*

$$\sigma_{sc}(H) \cap (\underline{m}, \bar{m}) = \phi.$$

Here,

$$\underline{m} = \lim\inf m(t) \quad and \quad \bar{m} = \lim\sup m(t)$$

Proof. The proof is analysed both for accumulation of eigenvalues and boundedness of the M-matrix. $\sigma_{sc}(H)$ cannot lie within the interval $(\underline{m}, \bar{m})$ since if X is an open subset of $\mathbb{C}$ such that $(\underline{m}, \bar{m}) \subset X$, then we may assume that for $z \in X$,the solutions $y_j(t, z)$, j = 1,..., 4, of $(L - z)y$ analytically depend on z such that for $z \in X$ with $Imz > 0$, then $\det L = (2, 2)$ and the point spectrum has no accumulation point within$(\underline{m}, \bar{m})$.The solutions $y_j(t, z)$, form the fundamental system $Y(t, z)$ of (0.0.1) since otherwise there would exist a solution which is in the domain of self-adjoint extension operator, implying that z is an eigenvalue. By analyticity, it follows that $y_j(t, z)$, j=1, ...,4, form a fundamental system for all $z \in X$, with possible exception for at most countably many

8

points which cannot accumulate in X.

Finally we show that the solutions that lose their square summability as $Imz \to 0^+$ cannot contribute to singular continuous spectrum. For $z \in \mathbb{R}$, those eigenvalues $,\lambda$, with $|\lambda| < 1$ and $|\lambda| > 1$ will lead to eigenfunctions which are z-uniformly square summable and z-uniformly non-square summable respectively and hence discrete spectrum at most. But if $Imz > 0, z \in \mathbb{C}$, then as $Imz \to 0^+$, some of the eigenfunctions from eigenvalues $\lambda, |\lambda| = 1$, lose their square summability, but since the domain of H is defined by only those eigenfunctions that are z-uniformly square summable, we need to show that $ImM(z)$ exists finitely and is bounded.

Now let $F(.,z)$ be 2 by 4 system of square summable solutions which satisfy α-boundary conditions at 0 and define the M-matrix $M(z)$ (see C.Remling [?])

$$< F(.,z), F(.,z') > (\bar{z} - z') = M^*(z) - M(z'), \qquad (0.0.10)$$

whose discrete version is given in [7]. Then for $z = z + i\eta, z_0 \in \mathbb{R}$ we have

$$ImM(z_0) = \lim_{\eta \to 0^+} \eta < F(., z_0 + i\eta); F(., z_0 + i\eta) > .$$

Assume the α-boundary conditions does not give rise to a bound state since otherwise $ImM(z_0)$ will exist boundedly, then the above limit exist finitely. To see this we use the two eigensolutions given by the eigenvalues λ such that $|\lambda| < 1$ even as $Imz \to 0^+$. In this case,

$$y_j(t, z) \simeq c_{jk}\lambda^t$$

and

$$y_i \simeq c_{ik}\lambda^t,$$

where c_{ik} and c_{jk} appropriate eigenvectors and are bounded. Then by Cauchy-Swartz inequality, we obtain

$$ImM(z_0) = \lim_{\eta \to 0^+} \eta| < y_i(t, \eta), y_j(t, \eta) > | \leq$$

$$\lim_{\eta \to 0^+} \eta (\sum_{k=1}^{4} |c_{ik}|^2 |\lambda_i^t(\eta)|^2)^{\frac{1}{2}} (\sum_{k=1}^{4} |c_{jk}|^2 |\lambda_j^t(\eta)|^2)^{\frac{1}{2}}.$$

The term on the right hand side is bounded absolutely as $t \to \infty$ since

9

$|\lambda_j(\eta)|, |\lambda_i(\eta)| < 1$. Consequently, $ImM(z_0)$ is non-trivial. This shows that the spectrum of H has no singular continuous part. $\qquad\square$

The interval $(\underline{m}, \overline{m})$ is not necessarily empty since if we assume that $m(t) = sin\frac{(t+1)}{2}\Pi$, then $\underline{m} = -1$ and $\overline{m} = 1$, hence $(\underline{m}, \overline{m}) = (-1, 1)$ yet $m(t)$ is bounded for all $t \in \mathbb{N}$.
Suppose that m, p are bounded with $q = r = 0$, then $(0.0.8)$ becomes a biquadratic whose zeros can be solved explicitly. Therefore under various asymptotic conditions we obtain the following result which has been proved in Agure, Ambogo and Nyamwala [1] and we provide proof for completeness.

Theorem 0.7. *Let m and p be bounded and suppose all the necessary and sufficient conditions for asymptotic summation are satisfied, then*

(i) *If $q = r = 0$ and $p^2 < 4(m - z)$ then $def L = (2, 2)$ and $\sigma(H)$ is pure discrete.*

(ii) *Assume all the coefficients are almost constant and that the limiting characteristic polynomial has $2l$ eigenvalues of absolute value one $(0 \le l \le 2)$, then the self-adjoint extension operator H has no singular continuous spectrum and $\sigma_{ac}(H)$ agrees with that of the constant coefficient limiting operator and has spectral multiplicity of l.*

Proof.   (i) Assume that $r(t) = q(t) = 0$ for all $t \in \mathbb{N}$ and the other coefficients bounded then the polynomial is a well known biquadratic polynomial that can be solved explicitly. Thus if $p^2 < 4(m - z)$, the discriminant of the polynomial is less than zero and hence the roots have non-zero imaginary parts. These roots are in complex conjugate pairs. Assume these roots are of the form $\alpha_j \pm \beta_j$, $j = 1, 2$. Using analysis given in [3], the two roots with $\beta_j > 0$ will lead to eigensolutions that are z-uniformly square summable while the two roots with $\beta_j < 0$ will lead to z-uniformly non-square summable eigensolutions. Thus def $L = (2, 2)$ and the spectrum is discrete at most.

(ii) If the coefficients are almost constant, then those roots λ of

$$\mathcal{P}(\lambda, t, z) \quad \text{such that} \quad |\lambda| > 1$$

lead to solutions that are z-uniformly non-square summable while the roots $|\lambda^{-1}| < 1$ lead to z-uniformly square summable solutions.

10

Therefore, it is the roots λ such that $|\lambda| = 1$, that lead to eigenfunctions of which half of their number lose their square summability as $Imz \to 0^+$. The eigenfunctions that lose their square summability as $Imz \to 0^+$ contribute to absolutely continuous spectrum.

Invoking the results of [2], the absolutely continuous spectrum of H coincides with that of the constant coefficients limiting operator and of spectral multiplicity equal to the number of eigenfunctions that lose their square summability as $Imz \to 0^+$.

$\square$

The following example confirms the results of the above two theorems. Before we give the example, we state the following lemma which is from classical linear algebra.

Lemma 0.8. *If λ and $\bar{\lambda}^{-1}$ are roots of the characteristic polynomial $\rho(\lambda, t, z)$ and assume that $\zeta = \lambda + \bar{\lambda}^{-1}$, then $|\zeta| \leq 2$ is only possible if ζ is real otherwise $|\zeta| > 2$.*

Lemma 0.16 implies that we can obtain $|\lambda| \simeq |\bar{\lambda}^{-1}| \simeq 1$ only if $\lambda + \bar{\lambda}^{-1}$ is real otherwise we will have $|\lambda| > 1$ and $|\bar{\lambda}^{-1}| < 1$.

Example 0.9. Let L be a fourth order difference operator generated by a difference equation of the form

$$\triangle^4 y(t-2) - \triangle\{(c_p + t^{\beta_p})\triangle y(t-1)\} + (c_m + t^{\beta_m})y(t) = zy(t)$$

where β_m, $\beta_p < 0$ and $c_p, c_m > 0$ are constants.
Then one can easily convert the above difference equation into its first order system using quasi-differences. Here as $t \to \infty$, then

$$c_p + t^{\beta_p} \to c_p, \quad \text{while} \quad c_m + t^{\beta_m} \to c_m.$$

The characteristic polynomial $\mathcal{P}(\lambda, t, z)$ multiplied by λ^{-2} becomes

$$(1-\lambda)^2(1-\lambda^{-1})^2 + (c_p + t^{\beta_p})(1-\lambda)(1-\lambda^{-1}) + c_m + t^{\beta_m} - z = 0$$

Now let $\lambda + \lambda^{-1} = \zeta$ so that we have

$$(2-\zeta)^2 + (2-\zeta)(c_p + t^{\beta_p}) + (c_m + t^{\beta_m} - z) = 0$$

11

Solving for ζ by absorbing $t^{\beta_p}, t^{\beta_m} - z$ into c_p and c_m respectively we get

$$\zeta_+ = 2 + \frac{c_p}{2} + \{\frac{c_p^2}{4} - c_m\}^{\frac{1}{2}}$$

$$\zeta_- = 2 + \frac{c_p}{2} - \{\frac{c_p^2}{4} - c_m\}^{\frac{1}{2}}.$$

Thus we have two broad cases to consider.

(a) $\beta_p < \beta_m < 0$. Then ζ_+ and ζ_- will be in complex conjugate pairs with non-zero imaginary parts. Applying the results of Lemma 0.16 ,both ζ_+ and ζ_- have absolute value greater than 2, hence each contribute (1,1) to deficiency index and the eigensolutions that are square summable are z-uniformly square summable. Hence $def L = (2, 2)$ and $\sigma(H)$ is discrete at most.

(b) $\beta_m < \beta_p < 0$. This can be split into three cases as follows

(i) $|\zeta_-| \leq 2, |\zeta_+| > 2$ then the expansion of

$$\zeta_- = 2 + \frac{c_p}{2} - \{\frac{c_p^2}{4} - c_m\}^{\frac{1}{2}} \approx 2 + \frac{c_p}{2} - \frac{c_p}{2}\{1 - \frac{2c_m}{c_p^2} + ..\}.$$

Thus after two diagonalisations, we need that the correction term be summable. The term affected by this, is that associated to the spectral parameter z which is $\frac{c_m}{c_p^2}$.Hence

$$\triangle^2(\frac{c_m}{c_p^2}) \approx \triangle^2(t^{\beta_m} - 2\beta_p) \approx O(t^{\beta_m - 2\beta_p - 2}).$$

Therefore if $\beta_m - 2\beta_p < 1$ then $def L = (3, 3)$ and $\sigma(H)$ is pure discrete. But if $\beta_m - 2\beta_p > 1$ then $\text{def} L = (2, 2)$ and $\sigma_{ac}(H) \subset [c_m, 16 + 4c_p + c_m]$ and has a spectral multiplicity of 1

(ii) If we assume $|\zeta_+| \leq 2$ and $|\zeta_-| > 2$, we obtain similar results as in (i) above

(iii) Suppose $|\zeta_+|, |\zeta_-| \leq 2$, this is possible since c_p and c_m can be chosen appropriately. Then if $\beta_m - 2\beta_p < 1$,$def L = (4, 4)$ and $\sigma(H)$ is pure discrete while if $\beta_m - 2\beta_p > 1$ then $def L = (2, 2)$ and $\sigma(H) \subset [c_m, 16 + 4c_p + c_m]$ of spectral multiplicity 2.

References

[1] **Agure J. O., Ambogo D. O. and Nyamwala F. O.**, Deficiency Indices and Spectrum of Fourth Order Difference Equations with Unbounded Coefficients, *Math.Nach*, 286(2013), 323-339.

[2] **Behncke H.** ,*Spectral theory of Higher Order Operators : Proc. of London Math* soc., (2006).

[3] **Behncke H. and Nyamwala F.O.**, Spectral Theory of Difference Operators With Almost Constant Coefficients II. *J.Difference Equations and Applications*, 17(5) (2011), 821-829.

[4] **Behncke H. and Nyamwala F.O.**, Spectral Theory of Difference Operators With Almost Constant Coefficients, *J.Difference Equations and Applications*. 17(5) (2011) 677-695.

[5] **Benzaid Z. and Lutz D.A.**, Asymptotic representation of solutions of perturbed Systems of Linear Difference Equations., Studies in Applied Math .77 (1987),195-221.

[6] **Hinton D.B and Shaw J.K**, *On the Titchmarsh-Weyl M(λ) Functions for Linear Hamiltonian Systems, J.Differential Equations* 40 (1981), 316-342.

[7] **Shi Y.**, Weyl-Titchmarsh *Theory for a Class of Discrete Linear Hamiltonian Systems .,Linear Algebra And its Appl.* 416 (2006), 452-519.

YOUR KNOWLEDGE HAS VALUE

- We will publish your bachelor's and master's thesis, essays and papers

- Your own eBook and book - sold worldwide in all relevant shops

- Earn money with each sale

Upload your text at www.GRIN.com and publish for free